BIOHACKING

TIPS FOR UNDERSTANDING ALL THINGS

KATE .P

Contents

CHAPTER ONE 3
INTRODUCTION 3
   What Biohacking Means 8
   Comprehending Biohacking 13
   Resources and Methods for Biohacking 17
CHAPTER TWO 18
   Uses for Biohacking 24
   Hazards and Ethical Issues 31
CHAPTER THREE 37
   Communities and Culture of Biohacking 37
   The Landscape of Law and Regulation 43
   Summary 50
THE END 54

CHAPTER ONE

INTRODUCTION

The phrase "biohacking" refers to a broad range of techniques that combine technology, lifestyle changes, and self-experimentation to maximize human performance, health, and well-being. It entails adopting a proactive and customized strategy to improve different facets of cognitive and physical function, frequently in order to reach peak performance, longevity, or general vitality.

Fundamentally, biohacking involves using state-of-the-art technologies and scientific understanding to manipulate the body's biological systems in order to accomplish specific goals. This can entail tactics for improving cognition as well as stress management, exercise routines, sleep optimization, nutritional supplements, and stress management.

Quantified self-tracking tools and biomarkers are frequently used by biohackers to track physiological indicators, collect information, and make well-informed choices regarding their performance and well-being. Biohackers want to maximize their biology and realize their maximum potential by examining this data and trying various interventions.

A wide variety of techniques are included in the field of biohacking, such as:

Nutritional optimization is the process of maximizing nutrient intake and metabolic health through experimenting with customized food plans, fasting schedules, and supplementation plans.

Lifestyle Modifications: Taking up routines and habits that support mental toughness, emotional health, and physical fitness, such as frequent exercise, mindfulness exercises, and stress management methods.

Sleep enhancement is the application of techniques to increase the quantity and quality of sleep, such as employing sleep monitoring devices, creating nighttime routines, and enhancing the sleep environment.

Cognitive Enhancement: Examining strategies to improve mental capacity and cognitive function, including brain training activities, nootropic supplements, and neurofeedback procedures.

Biometric monitoring is the process of keeping an eye on physiological factors including heart rate, sleep patterns, exercise levels, and biomarkers of performance and health using wearable technology and health monitors.

Utilizing genetic testing services to determine a person's genetic predispositions and customizing interventions for the best possible health outcomes is known as genetic testing and personalized medicine.

Participating in citizen science programs, DIY biology projects, and cooperative biohacking networks can help to democratize access to biotechnology and foster innovation in health and wellness.

Although biohacking has the potential to improve human performance and wellbeing, there are moral, security, and legal issues to be addressed. Critics point to the necessity for responsible oversight and informed permission, the possibility of hazards associated with self-experimentation, and the lack of scientific rigor in some biohacking techniques.

Achieving a balance between innovation and safety, ethical considerations and evidence-based methods is crucial as biohacking gains momentum and continues to advance. Biohacking has the ability to enable people to take charge of their health and open up new avenues for human potential with cautious experimentation, wise decision-making, and a dedication to ethical ideals.

What Biohacking Means

The term "biohacking" describes the activity of altering and improving living things including humans by means of biology, biotechnology, and other scientific fields in order to accomplish certain goals. It includes a wide range of endeavors intended to enhance biological systems, advance human potential, and enhance health and well-being.

Fundamentally, biohacking entails breaking into living things' biological systems in order to modify their features, traits, or behavior. Genetic engineering, biochemical control, cognitive improvement, and lifestyle changes are a few examples of this. To study, experiment, and create in the realm of biology, biohackers frequently combine scientific knowledge, technology instruments, and do-it-yourself (DIY) methods.

A biohacker may use a variety of techniques, such as:

DIY Biology: The study of biology by individuals and groups conducting experiments, investigating scientific issues, and coming up with creative solutions outside of conventional laboratory settings.

Biomedical hacking is the practice of monitoring, diagnosing, and treating medical disorders using biotechnology, medical equipment, and digital health applications; this is frequently done with an emphasis on self-directed healthcare and personalized treatment.

Experimenting with diet, supplements, and lifestyle changes to maximize nutrition, metabolism, and general health and performance is known as "nutritional biohacking."

Cognitive Enhancement: Examining strategies like brain training, nootropics, and neurofeedback to improve memory, mental function, and performance.

Body modification is the use of bioengineering methods to alter the human body in order to improve it aesthetically, functionally, or sensory-wise. Examples of these methods include implanting microchips or improving prosthetics.

Quantified Self: Using wearable technology, sensors, and self-tracking tools to measure and analyze personal health, fitness, and well-being data in order to guide behavior modification and decision-making.

Although biohacking has the potential to improve human performance and health, advance scientific understanding, and empower individuals, it also brings up moral, legal, and safety issues. Critics point to the necessity for informed permission, the possible dangers of unsupervised experimentation, and the ramifications of changing biological systems without fully appreciating the long-term effects.

Striking a balance between innovation and accountability is crucial as the area of biohacking develops, encouraging moral behavior, safety precautions, and evidence-based strategies. Biohacking has the potential to significantly improve healthcare, biotechnology, and human well-being while addressing ethical and societal issues if it is properly supervised, educated, and collaboratively undertaken.

Comprehending Biohacking

To comprehend biohacking, one must grasp the idea of optimizing human performance, health, and well-being by utilizing biology, technology, and scientific knowledge. Fundamentally, biohacking entails taking a proactive, individualized approach to improving many facets of cognitive and physical function, frequently by use of data-driven techniques and self-experimentation.

The term "biohacking" refers to a wide variety of methods and approaches, such as:

Nutritional optimization is the process of experimenting with food interventions to improve metabolism, energy levels, and general health. These interventions include fasting protocols, macronutrient manipulation, and customized nutrition regimens.

Lifestyle Modifications: Putting into practice lifestyle adjustments to promote the best possible physical and mental well-being, such as regular exercise, stress reduction methods, and good sleep hygiene.

Cognitive Enhancement: Examining methods to improve memory, mental acuity, and cognitive function by using activities for brain training, nootropic supplements, and mindfulness exercises.

Biometric monitoring is the process of collecting data for well-informed decision-making by monitoring physiological factors including heart rate variability, sleep patterns, and stress levels using wearable technology, health trackers, and biofeedback tools.

Genetic Analysis: Assessing a person's genetic predispositions and customizing strategies for optimal performance and individualized health through the use of genetic testing services.

DIY Biology Projects: Using citizen science programs and do-it-yourself biology experiments to investigate scientific issues, carry out investigations, and create novel answers outside of conventional laboratory settings.

Engaging in biohacking networks, online forums, and cooperative projects as a means of exchanging ideas, exchanging knowledge, and providing mutual support to individuals pursuing health and performance optimization.

Recognizing the safety, legal, and ethical issues related to independent biological system alteration and experimentation is another necessary step in understanding biohacking. Respecting individual autonomy, encouraging ethical behavior, and placing a high priority on safety, informed consent, and evidence-based procedures are all components of responsible biohacking.

To optimize their biology, unleash their potential, and improve their quality of life, people can investigate the concepts and methods of biohacking by adopting an attitude of curiosity, experimentation, and self-empowerment. To guarantee that treatments are secure, efficient, and in line with personal objectives and values, biohacking must be approached cautiously, critically, and with a dedication to ethical standards.

Resources and Methods for Biohacking

Using a variety of instruments and methods, biohacking aims to maximize human biology, well-being, and productivity. These tools cover a broad spectrum of scientific techniques, technological innovations, and lifestyle choices meant to improve mental, emotional, and physical health.

CHAPTER TWO

The following are some typical instruments and methods used in biohacking:

Wearable Technology: Wearable devices track physiological factors like heart rate, sleep patterns, activity levels, and stress reactions. Examples of these devices include fitness trackers, smartwatches, and biometric sensors. These gadgets offer real-time feedback and data analysis so users can monitor their progress, spot patterns, and decide which lifestyle adjustments to make.

Applications for Biometric Monitoring: Wearable technology and other sources of biometric data can be tracked and analyzed by users through mobile applications and software platforms. For the purpose of maximizing performance and wellness, these apps include functions including goal-setting, trend analysis, data visualization, and personalized recommendations.

Genetic Testing Services: DNA analysis kits that reveal a person's unique genetic predispositions regarding nutrition, fitness, health, and ancestry are available from direct-to-consumer genetic testing companies. hereditary testing can be used to optimize lifestyle decisions based on hereditary variables, identify possible dangers, and inform individualized interventions.

Nutritional Tracking Tools: Users can monitor the balance of macronutrients, keep track of their food intake, and examine the nutritional value of various foods with the help of mobile apps and software platforms. These resources assist people in identifying eating trends, optimizing nutrition, and making well-informed decisions to support their performance and health objectives.

Biofeedback Devices: Biofeedback devices provide information on stress levels, relaxation methods, and cognitive performance by monitoring physiological responses such as skin conductance, heart rate variability, and brainwave activity. Through biofeedback training, people can enhance their self-awareness and learn to control their physiological functions.

Nootropic Supplements: Nootropics are compounds that are said to boost memory, focus, and cognitive function. They are also referred to as smart medications or cognitive enhancers. Caffeine, omega-3 fatty acids, and herbal medicines like bacopa monnieri and ginkgo biloba are typical examples. In order to promote brain function and health, nootropic supplementation is frequently paired with lifestyle changes.

Brain Training Apps: Applications and software for cognitive training provide games and activities aimed at testing and enhancing mental skills including recall, focus, quick thinking, and memory. Over time, brain training can be utilized to sustain mental acuity and improve cognitive performance.

Sleep Tracking Apps and Devices: Sleep tracking apps and devices measure the length, pattern, and quality of sleep to assist people in identifying sleep problems, enhancing their sleep hygiene, and developing better sleeping habits. Interventions to improve general well-being and sleep quality can be informed by data from sleep tracking devices.

DIY Biology Kits: These kits come with everything needed to perform genetic engineering, biohacking, and biological studies in community labs or at home. With the help of these kits, hobbyist researchers and biohackers can investigate biological systems, carry out experiments, and create new things outside of conventional lab environments.

Biohacking Communities and Resources: Biohackers can exchange ideas, share expertise, and work together on projects through social media groups, online forums, and community meetups. For those who are interested in biohacking, these groups provide materials, networking opportunities, and support.

It's crucial to remember that, despite the potential for improving performance and health, biohacking equipment and methods can also have drawbacks and hazards. To guarantee that treatments are secure, efficient, and in line with personal objectives and values, responsible biohacking entails making well-informed decisions, taking safety precautions, and abiding by ethical standards.

Uses for Biohacking

Numerous fields, including healthcare and medicine, personal growth, athletic performance, and scientific study, have found uses for biohacking. The novel strategies and methods used in biohacking have the potential to enhance human biology, enhance health outcomes, and further scientific understanding. Here are a few significant biohacking applications:

Personalized Medicine: By using lifestyle modifications, genetic testing, and biomarker monitoring to customize treatments and preventive measures to each patient's unique needs, biohacking makes personalized approaches to healthcare and medicine possible. Based on genetic predispositions, lifestyle choices, and environmental influences, personalized medicine seeks to maximize therapeutic efficacy, reduce adverse effects, and enhance patient outcomes.

Health Optimization: People can optimize their health, prevent chronic diseases, and improve their general well-being by using biohacking techniques like nutritional optimization, lifestyle adjustments, and biometric monitoring. Through the monitoring of biomarkers, data analysis, and well-informed lifestyle decisions, people can take charge of their health and enhance their quality of life.

Enhancement of Performance: Biohacking techniques are applied to improve one's physical, mental, and emotional capabilities in a variety of contexts, such as academics, professional pursuits, and sports. Strategies including stress reduction, cognitive training, and nootropic supplementation help people perform at their best when faced with challenges.

Longevity and Anti-Aging: By tackling the fundamental biological pathways linked to aging, biohacking techniques seek to both promote and slow down the aging process. To encourage healthy aging and prolong longevity, interventions like calorie restriction, intermittent fasting, and antioxidant supplements target aspects linked to oxidative stress, inflammation, and cellular aging.

Enhanced Regeneration and Recovery: Following an illness, injury, or physically demanding activity, biohacking techniques promote enhanced regeneration and a quicker recovery time. Techniques like cryotherapy, cold exposure, and regenerative therapies improve muscle damage and exhaustion from exercise-induced weariness while lowering inflammation and enhancing tissue healing.

Bioinformatics and Data Analytics: Biohacking is the application of bioinformatics tools and data analytics approaches to the analysis of massive datasets pertaining to biology and health. In order to guide individualized interventions and scientific breakthroughs, researchers can find patterns, correlations, and insights by mining data from sources including genetic databases, electronic health records, and wearable technology.

DIY Biology and Citizen Science: To democratize access to scientific knowledge and foster biotechnology innovation, biohacking promotes involvement in do-it-yourself biology projects and citizen science efforts. Online platforms, community biohacking spaces, and do-it-yourself biology labs encourage amateur scientists, amateurs, and enthusiasts to collaborate, share knowledge, and conduct experiments.

Sustainability and Environmental Monitoring: Ecological restoration, environmental monitoring, and sustainability initiatives are all included in the biohacking scope, which goes beyond human biology. Monitoring environmental health, reducing pollution, and fostering ecological resilience are made possible through home-made environmental sensors, biomimetic architecture, and bioremediation methods.

Biotechnology Innovations: By promoting experimentation, prototyping, and investigation of new applications and technologies, biohacking promotes biotechnology innovation. Bioengineering initiatives, bioart installations, and do-it-yourself biotech projects push the limits of scientific innovation and technical progress.

Public Engagement and Education: Biohacking encourages scientific literacy, public science education, and a passion for STEM (science, technology, engineering, and mathematics) fields. Opportunities for experiential learning, investigation, and discovery are offered by biohacking workshops, outreach initiatives, and educational materials.

In general, biohacking has many different and complex applications in the fields of healthcare, research, education, and innovation. Individuals and communities can explore new avenues in health, performance, and scientific research by utilizing the concepts and methods of biohacking, leading to improvements that benefit society as a whole.

Hazards and Ethical Issues

While biohacking has the potential to advance scientific understanding, boost performance, and improve health, it also carries a number of hazards and ethical issues that need to be properly considered. Among them are:

Safety Risks: Individuals may be at risk for injury from some biohacking interventions, especially those that include do-it-yourself biology projects or self-experimentation with illegal chemicals. Biohacking experiments may cause harm or unexpected results if there is inadequate supervision, experience, or training.

Health Concerns: There may be unidentified health effects or interactions from biohacking techniques such supplement self-administration, genetic alteration, or testing out novel treatments. If biohacking is done without solid scientific backing and medical supervision, participants run the risk of endangering their health.

Data security and privacy: Gathering and disseminating private biological, genetic, and health-related information is a common practice in biohacking. Concerns have been raised over the security and privacy of this data, including the possibility of data breaches, unauthorized access, and third parties misusing personal information.

Informed Consent: The ethical precepts of informed consent, voluntary involvement, and transparency should be upheld in biohacking experiments, particularly those involving human subjects. Before taking part in trials or research projects, people must be made aware of the possible risks, advantages, and uncertainties related to biohacking treatments.

Regulatory Compliance: The regulatory bodies that oversee biotechnology, nutritional supplements, genetic testing, medical devices, and genetic testing may have jurisdiction over biohacking operations. Ensuring lawful and ethical conduct necessitates compliance with regulatory regulations, which includes getting the necessary permissions, licenses, and permits.

Equity and Access: Disparities in access to resources, knowledge, and technologies are among the issues surrounding equity and access in the field of biohacking. Practices related to biohacking could make inequality worse, especially if they are limited to affluent people or groups.

Misinformation and pseudoscience: There are a wide range of people in the biohacking community, each with a different degree of scientific knowledge and proficiency. In the discourse around biohacking, there exists a potential for disinformation, pseudoscience, and unsupported claims to cause confusion, injury, and misunderstanding.

Social and Cultural Implications: Biohacking interventions could raise issues related to identity, authenticity, and society norms, among other more general social and cultural issues. The adoption of biohacking by society, its effects, and its implications for social cohesiveness and personal well-being raise ethical questions.

Dual-Use Concerns: Biohacking techniques and technologies have the potential to be applied for both good and bad reasons, or for malevolent or harmful intents. Reducing the possibility of biohacking inventions being abused, misused, or having unexpected repercussions is one ethical consideration.

Environmental Impact: It is important to carefully consider and mitigate any potential environmental effects of certain biohacking techniques, such as genetic engineering and bioremediation projects. Unintentional ecological consequences, ecosystem contamination, and biodiversity disturbance are among the risks.

A multidisciplinary strategy combining cooperation between scientists, legislators, ethicists, regulators, and community stakeholders is needed to address these hazards and ethical issues. We can guarantee that biohacking occurs in a safe, morally sound, and equitable manner while optimizing its potential advantages for people and society by encouraging responsible behavior, moral supervision, and open communication.

CHAPTER THREE

Communities and Culture of Biohacking

The community and culture surrounding biohacking are made up of a wide range of people, including hobbyists, scientists, and business owners, who are all interested in maximizing human biology, health, and performance through do-it-yourself experimentation, creativity, and teamwork. These communities offer forums for knowledge sharing, idea exchange, and mutual support in the pursuit of biohacking objectives. They also cultivate a culture of empowerment, curiosity, and discovery. The following are some salient features of biohacking societies and cultures:

Diversity: People from a variety of backgrounds, including scientists, engineers, medical professionals, tech enthusiasts, and hobbyists, make up the biohacking groups. The variety of viewpoints and areas of expertise encourages multidisciplinary cooperation and creativity in biohacking projects.

Cooperation: To address challenging issues and enhance scientific understanding, biohackers frequently work together on projects, experiments, and research endeavors. They do this by pooling resources, skills, and knowledge. Online discussion boards, community labs, and collaborative platforms offer venues for networking, idea exchange, and group education.

Openness and Transparency: The culture of biohacking places a high importance on information sharing, openness, and transparency. Participants are encouraged to record their methods, findings, and experiments for the good of the community. Peer review, repeatability, and accessibility are encouraged in biohacking research and innovation by open scientific principles.

Innovation and DIY Spirit: Biohacking is a do-it-yourself (DIY) movement that encourages people to explore on their own and learn on their own by taking charge of their health and wellbeing. Maker spaces, citizen science programs, and do-it-yourself biology projects open up biotechnology to the public and encourage local ingenuity.

Education and Skill Development: Workshops, training courses, and instructional materials are all part of the biohacking communities' efforts to provide people with the information and resources they need to engage in biohacking in a responsible and safe manner. Participants gain competence and confidence through practical learning opportunities and mentorship.

Ethical Behavior and Responsibility: To protect people's health and the objectivity of scientific study, the biohacking community places a strong emphasis on ethical behavior, responsible experimentation, and adherence to safety procedures. Ethical dialogues, protocols, and social standards encourage moral judgment and responsibility in biohacking activities.

Empowerment and Self-Optimization: Biohacking encourages people to take charge of their own health, biology, and personal growth by giving participants a sense of empowerment and agency. In order to reach their maximum potential, people use biohacking interventions to maximize their physical, cognitive, and emotional well-being.

Community Support and Networking: Members of biohacking groups benefit from one another's support, encouragement, and friendship, which fosters a sense of unity and belonging among people who have similar interests. In biohacking initiatives, networking, collaboration, and mutual help are facilitated by peer support networks, online forums, and community events.

Public Engagement and Advocacy: In order to increase scientific literacy, spread understanding of biohacking principles, and support moral, just, and inclusive approaches to biotechnology and human enhancement, biohacking communities participate in public outreach, advocacy, and education campaigns.

Cultural Impact and Social Change: The culture of biohacking has a significant impact on attitudes, customs, and laws pertaining to technology, health, and human enhancement. Biohacking encourages cultural shifts toward customized, proactive methods to self-care and personal development by questioning traditional ideas of health and wellness.

All things considered, the culture and communities around biohacking are essential for stimulating creativity, encouraging teamwork, and encouraging ethical use of biotechnology for human development. Biohacking communities enable people to explore new frontiers in health, performance, and self-discovery while advancing moral principles and the welfare of society by creating an open and encouraging environment.

The Landscape of Law and Regulation

The legal and regulatory environment that surrounds biohacking is complicated and varies based on a number of variables, including the country, the type of actions being undertaken, and the desired results. The laws and rules governing fields like healthcare, biotechnology, genetics, consumer goods, and intellectual property may cross paths with biohacking operations. The following are some significant facets of the biohacking legal and regulatory environment:

Healthcare Authorities: The Food and Drug Administration (FDA) in the United States and the European Medicines Agency (EMA) in Europe may impose regulations on biohacking operations involving medical equipment, diagnostic tests, or therapeutic treatments. Medical biohacking procedures must adhere to regulatory regulations for efficacy, safety, and quality assurance.

Biotechnology rules: Genetic engineering, gene editing, and other biohacking techniques that modify biological materials could be subject to biotechnology rules. Biohacking initiatives involving genetically modified organisms (GMOs) or the discharge of modified organisms into the environment may be supervised by regulatory organizations like the Environmental Protection Agency (EPA) and the United States Department of Agriculture (USDA).

Intellectual Property Laws: Patents, copyrights, and trademarks are only a few examples of the intellectual property laws that may protect biohacking ideas, inventions, and discoveries.

Biohackers can protect their inventions, manage commercialization, and stop illegal use or duplication of their intellectual property with the help of intellectual property rights.

Laws Protecting Consumers: Biohacking goods, services, and technology sold to consumers may be governed by laws and rules protecting consumers, including those pertaining to product safety, advertising, and labeling. Adherence to consumer protection legislation guarantees that biohacking products satisfy quality and safety criteria and furnish consumers with precise information.

Data Privacy and Security Laws: Biohacking operations that gather, store, or distribute genetic or personal health data may be governed by data privacy and security laws. Data protection, confidentiality, and informed consent are mandated by laws like the General Data Protection Regulation (GDPR) in Europe and the Health Insurance Portability and Accountability Act (HIPAA) in the United States.

Institutional Review Boards (IRBs) or ethics committees may oversee biohacking experiments involving humans or animals and are responsible for adhering to ethical rules. Ethical biohacking research must adhere to ethical standards such as informed consent, voluntary involvement, and participant risk minimization.

Import and Export rules: Customs rules and export control acts may apply to biohacking operations that include the import, export, or movement of biological materials, tools, or technologies. Transporting biohacking materials across borders legally and securely requires adherence to import/export regulations, license, and paperwork.

Professional Standards and Codes of Conduct: Professional associations or licensing boards may have created codes of conduct, disciplinary procedures, and professional standards for biohackers, especially those working in the healthcare or scientific research sectors. Sustaining professional credibility and trust necessitates adherence to ethics, honesty, and responsibility.

Public Health and Environmental Regulations: Biohacking initiatives that could have an impact on the environment or public health, like the release of genetically modified organisms, bioremediation, or ecological restoration, may be governed by laws. To reduce hazards and guarantee responsible behavior, adherence to laws controlling environmental risk assessment, biosecurity, and biosafety is crucial.

Emerging Policy Issues: As biohacking develops and becomes more well-known, legislators, regulators, and stakeholders must deal with new issues pertaining to biosecurity, dual-use concerns, biotechnology access, and the social effects of human improvement. In order to balance innovation with safety, ethics, and societal values, addressing these policy concerns calls for proactive involvement, stakeholder participation, and evidence-based policymaking.

Understanding relevant laws, rules, and standards is essential for navigating the legal and regulatory environment around biohacking. Proactive compliance initiatives, risk management techniques, and ethical concerns are also important. Policymakers, regulators, biohackers, and other stakeholders must work together to create regulatory frameworks that support ethical norms in biohacking methods, encourage innovation, and safeguard public health and safety.

Summary

To sum up, biohacking is a vibrant and multidimensional movement that includes a range of behaviors, groups, and cultural standards all aimed at maximizing human biology, health, and performance. Biohacking is a combination of DIY biology projects, individual experimentation, teamwork in research projects, and technological advancements that promotes empowerment, discovery, and creativity.

Because biohacking democratizes access to biotechnology, encourages individual agency, and cultivates a culture of self-optimization, it has the potential to transform scientific research, healthcare, and personal development. Biohackers are expanding the frontiers of science, improving human welfare, and igniting societal change via interdisciplinary cooperation, open science ideals, and moral behavior.

But biohacking also brings up significant moral, legal, and societal issues that need to be addressed in order to maintain moral norms, encourage responsible behavior, and safeguard public safety. Risks include privacy issues, safety issues, and difficulties with regulatory compliance highlight the necessity of rigorous oversight, well-informed decision-making, and adherence to moral standards in biohacking activities.

Going forward, biohackers, legislators, regulators, and stakeholders will shape the field through continued discussion, cooperation, and innovation. We can harness the potential of biohacking to improve human health, enhance quality of life, and address pressing societal challenges while upholding principles of safety, equity, and respect for individual autonomy by encouraging ethical conduct, evidence-based practices, and inclusive participation.

Biohacking presents opportunities and problems in this dynamic and developing subject that need for careful thought, responsible conduct, and a dedication to ethical principles. We can fully realize the potential of biohacking to build a more robust, resilient, and empowered future for people and communities everywhere by adopting these principles and cooperating.

THE END

www.ingramcontent.com/pod-product-compliance
Lightning Source LLC
Chambersburg PA
CBHW030053230526
45471CB00003B/1077